Math Mammoth
Factors & Factoring

By Maria Miller

Contents

Introduction

Math Mammoth Factors & Factoring is a worktext that covers typical number theory topics for upper elementary school: divisibility rules, finding factors of a given number, prime numbers, prime factorization, the greatest common factor, and the least common multiple. These topics are usually covered in grades 4-6.

The book begins with 4th grade topics, starting out with the concept of divisibility and the common divisibility rules. The lesson *Prime Numbers* brings out the concept of a prime number as one that is only divisible by 1 and by itself. Armed with the knowledge about divisibility, students now learn how to find factors of a given two-digit number in the lesson *Finding Factors*.

The following lesson, *Primes and Finding Factors*, is a review lesson about factors, primes, and divisibility rules, originally intended for beginning of 5th grade. You can use it as a review of the first four lessons of this book, or for students that have some knowledge of these topics already.

The next two lessons delve into prime factorization using a factor tree. The first lesson on the topic only uses two-digit numbers. In the latter, students are introduced the sieve of Eratosthenes and they factorize three-digit and larger numbers.

The last topics in the book are the greatest common factor (GCF), the least common multiple (LCM), and factoring sums, intended mainly for 6th grade. Factoring sums means writing a sum such as $42 + 18$ as $6(7 + 3)$. We first find a common factor of 42 and 18 (which is 6), and use that to factorize the sum. The lesson only deals with numeric examples, but it is actually preparing students for algebra, where the same process is done with variables; for example $x^2 + 2x$ is factored as $x(x + 2)$.

The book ends in a review lesson. Answers are appended.

You can find videos for the lessons in this on this page:

https://www.mathmammoth.com/videos/division_factors#factors

I wish you success in teaching math!

Maria Miller, the author

Helpful Resources on the Internet

We have compiled a list of external Internet resources that match the topics in this book. This list of links includes web pages that offer:

- **online practice** for concepts;

- online **games**, or occasionally, printable games;

- **animations** and interactive **illustrations** of math concepts;

- **articles** that teach a math concept.

We heartily recommend you take a look at the list. Many of our customers love using these resources to supplement the bookwork. You can use the resources as you see fit for extra practice, to illustrate a concept better, and even just for some fun. Enjoy!

https://l.mathmammoth.com/blue/factorsfactoring

Scan me

Divisibility

A number *n* is **divisible** by another number *m*, if the division $n \div m$ is exact (no remainder).

For example, $18 \div 3 = 6$, so 18 is divisible by 3.

Also, 18 is divisible by 6, because we can write the other division $18 \div 6 = 3$.
So, 18 is divisible by *both* 6 and 3. We say 6 and 3 are **divisors** of 18.

You can use long division to check if a number is divisible by another.

For example, $67 \div 4 = 16$, R3. There is a remainder, so 67 is *not* divisible by 4.

Also, from this we learn that neither 4 nor 16 are divisors of 67.

$$\begin{array}{r} 1\,6 \\ 4\,\overline{)\,6\,7} \\ -4 \\ \hline 2\,7 \\ -2\,4 \\ \hline 3 \end{array}$$

1. Divide and determine if the number is divisible by the other number.

a. $21 \div 3 =$ _____	b. $40 \div 6 =$ _____	c. $17 \div 5 =$ _____	d. $84 \div 7 =$ _____
Is 21 divisible by 3?	Is 40 divisible by 6?	Is 5 a divisor of 17?	Is 7 a factor of 84?

2. Answer the questions. You may need long division.

a. Is 98 divisible by 4?	b. Is 603 divisible by 7?	c. Is 3 a factor of 1,256?

In any multiplication, the numbers that are multiplied are called **factors** and the result is called a **product**.

factor	factor	product
7	× 6	= 42

For example, since $6 \times 7 = 42$, 6 and 7 are **factors** of 42.

From this multiplication fact we can write two divisions: $42 \div 6 = 7$ and $42 \div 7 = 6$. So, this means that 6 and 7 are also divisors of 42.

From this we can notice the following:

If a number is a factor of another number, it is also its divisor.

There is yet one more new word to learn that ties in with all of this: **multiple**.

We say **42 is a multiple of 6**, because 42 is some number times 6 (namely $\underline{7} \times 6$).

And of course 42 is also a multiple of 7, because 42 is some number times 7 (namely, $\underline{6} \times 7$)!

3. Fill in. We know that $8 \times 9 = 72$. So, 8 is a _____ of 72, and so is 9.

Also, 72 is a _____ of 8, and 72 is a _____ of 9.

And, 72 is _____ by 8 and by 9.

4. Fill in.

a. Is 5 a factor of 55? Yes, because ____ ÷ ____ = _____ .	**b.** Is 8 a divisor of 45? No, because ____ ÷ ____ = _____ .
c. Is 36 a multiple of 6? _____ , because ____ ÷ ____ = _____ .	**d.** Is 34 a multiple of 7? _____ , because ____ ÷ ____ = _____ .
e. Is 7 a factor of 46? _____ , because _____ .	**f.** Is 63 a multiple of 9? _____ , because _____ .

Multiples of 6 are all those numbers we get when we multiply 6 by other numbers. For example, we can multiply 0×6, 7×6, 11×6, 109×6, and so on. The resulting numbers are all multiples of six.

In fact, the skip-counting pattern of 6 gives us a list of multiples of 6:

0, 6, 12, 18, 24, 30, 36, 42, 48, 54, 60, 66, 72, 78, 84, and so on.

5. **a.** Make a list of multiples of 11, starting at 0 and continue at least to 154.

b. Make a list of multiples of 111, starting at 0. Continue as long as you can in this space!

Divisibility by 2

Numbers that are divisible by 2 are called **even** numbers.
Numbers that are NOT divisible by 2 are called **odd** numbers.

Even numbers end in 0, 2, 4, 6, or 8. Every second number is even.

Divisibility by 5

Numbers that end in 0 and 5 are divisible by 5.

For example, 10, 35, 720, and 3,675 are such numbers.

6. Mark an "x" if the number is divisible by 2 or by 5.

number	divisible by 2	by 5
750		
751		
752		
753		
754		

number	divisible by 2	by 5
755		
756		
757		
758		
759		

number	divisible by 2	by 5
760		
761		
762		
763		
764		

number	divisible by 2	by 5
765		
766		
767		
768		
769		

Divisibility by 10

Numbers that end in 0 are divisible by 10.

For example, 10, 60, 340, and 2,570 are such numbers.

7. Mark an "x" if the number is divisible by 2, by 5, or by 10.

number	divisible by 2	by 5	by 10
860			
861			
862			
863			
864			

number	divisible by 2	by 5	by 10
865			
866			
867			
868			
869			

number	divisible by 2	by 5	by 10
870			
871			
872			
873			
874			

If a number is divisible by 10, it ends in a zero, so it is ALSO divisible by ____ and ____.

8. **a.** Write a list of numbers that are divisible by 2, from 0 to 60.

_____

This is also a list of _____ of 2.

b. In the list above, *underline* those numbers that are divisible by 4.
What do you notice?

c. In the list above, *color* those numbers that are divisible by 6.
What do you notice?

d. Which numbers are divisible by both 4 and by 6?

9. **a.** Write a list of numbers that are divisible by 3, from 0 to 60.

_____

This is also a list of _____ of 3.

b. In the list above, *underline* those numbers that are divisible by 6.
What do you notice?

c. In the list above, *color* those numbers that are divisible by 9.
What do you notice?

10. Use the lists you made in (8) and (9). Find numbers that are divisible by *both* 2 and 9.

11. What number is a factor of every number?

12. Twenty is a multiple of 4. It is also a multiple of 5. It is also a multiple of four other numbers.
Which ones?

Who am I? (Hint: I am less than 50.) Mystery Number 38 25 31 99 47 101 9	*Who am I?* (Hint: I am less than 100.) Mystery Number 38 25 31 99 47 101 9
Divided by 9, I leave a remainder of 6. Divided by 4, I leave a remainder of 1. Divided by 10, I leave a remainder of 3.	I am a multiple of 3, 4, 5, and 6. I am a factor of 120. Divided by 7, I leave a remainder of 4.

Divisibility and Factors

Recall that in any multiplication, the numbers that are multiplied are called **factors** and the result is called a **product**.

factors   product

$6 \times 8 = 48$

(Even the multiplication "6 × 8" is called a product. You can call it the product written, whereas 48 is the product calculated or solved.)

We say that **6 is a factor of 48** -- because 6 times some whole number equals 48. Similarly, **8 is a factor of 48** (because 8 times some whole number equals 48).

From the multiplication fact above we can make two division facts: $48 \div 6 = 8$ and $48 \div 8 = 6$.

$48 \div 8 = 6$

dividend   quotient
divisor

These divisions are even divisions (no remainder). This means that both 6 and 8 are **divisors** of 48.

We also say that 48 is **divisible by** both 6 and 8.

If a number is a factor of another number, it is also its divisor.

1. How can you check whether a number is a factor of another number?
   For example, how would you check whether 7 is a factor of 623?

2. Answer the questions, and justify your answer.

   a. Is 8 a factor of 100? (Yes/No), because

   b. Is 9,896 divisible by 7?   (Yes/No), because

   c. Is 9 a divisor of 50?   (Yes/No), because

3. Mark said, "I know that 607 is divisible by 13, because its digits add up to 13 (6 + 0 + 7 = 13)."
   Is Mark correct? If not, prove to him the truth of the matter.

4. **a.** How can you tell, without calculations, whether the number 3 × 4 × 87 is divisible by 3?

   b. Is this number divisible by 12? Why or why not?

   c. Is 2 × 758 × 5 divisible by 10?

Easy divisibility rules (You should already know these.)

A number is **divisible by 2** if it ends in 0, 2, 4, 6, or 8. Such a number is even.

A number is **divisible by 5** if it ends in 0 or 5. For example, 395 is divisible by 5.

A number is **divisible by 10** if it ends in 0. For example, 56,930 is divisible by 10.

A number is **divisible by 100** if it ends in "00". For example, 450,000 is divisible by 100.

A number is **divisible by 1000** if it ends in "000". For example, 450,000 is divisible by 1000.

Example 1. Is 2 × 3 × 17 divisible by 10?

2 × 3 × 17 is 6 × 17. Imagine doing this multiplication in columns. What will be the *last* digit of the answer?

The answer will end in 2, because 6 × 7 = 42 ends in 2. Therefore, the number is not divisible by 10.

---

5. Mark an "x" if the number is divisible by 2, 5, 10, 100, or 1,000.

Divisible by	2	5	10	100	1000
825					
400					
332					

Divisible by	2	5	10	100	1000
600,200					
56,000					
307,995					

6. Answer. In each case, explain why or why not it is so (justify your answer).

   a. Is 6 × 28 divisible by 5?

   b. Is 3 × 794 divisible by 10?

   c. Is 3 × 3 × 3 × 3 × 3 divisible by 2?

   d. Is 2 × 15 × 2 × 7 divisible by 4?

Divisibility rules for 3, 6, and 9

A number is divisible by 3 if the sum of its digits is divisible by 3.

Example 2. To check if 93,025 is divisible by 3, add its digits: $9 + 3 + 0 + 2 + 5 = 19$.
Since 19 is *not* divisible by 3, neither is 93,025. (In fact, $93,025 \div 3 = 31,008$ R1.)

A number is divisible by 6 if it is divisible by both 2 and 3.

Example 3. The number 756 is divisible by 3, because the sum of its digits is 18, which *is* divisible by 3. Also, it is an even number so it's divisible by 2. Therefore, it is also divisible by 6. (In fact, $756 \div 6 = 126$.)

A number is divisible by 9 if the sum of its digits is divisible by 9.

Example 4. To check if 105,642 is divisible by 9, add its digits: $1 + 0 + 5 + 6 + 4 + 2 = 18$.
Since 18 *is* divisible by 9, so is 105,642.

7. Tell whether these numbers are divisible by 3.
   If yes, divide the number by 3 (long division).

   a. 539

   b. 43,719

   c. 9,032

Tip: In adding the digits, you can *omit* any digits that are divisible by 3 (namely 3, 6, and 9). For example, to check whether 99,378 is divisible by 3, just add $7 + 8 = 15$ and omit 9, 9, and 3. Since 15 is divisible by 3, so is 99,378.

8. Change one of the digits in the number 238,882 so
   that the number is divisible by 3, but *not* divisible by 2.

9. Tell whether these numbers are divisible by 9.
   If yes, divide the number by 9 (long division).

   a. 888

   b. 576

   c. 44,082

10. Mark an "x" if the number is divisible by 2, 3, 5, 6, or 9.

Divisible by	2	3	5	6	9
589					
558					

Divisible by	2	3	5	6	9
495					
3,594					

More on Divisibility

If you know that a number is divisible by some number n, then you can skip-count by n to find more numbers that are also divisible by n.

Example 1. Since 100 is divisible by 4, then, $100 - 4 = $ **96** is also divisible by 4. We can skip-count by fours from 100—up or down—to get a list of numbers that are divisible by 4:

100, 96, 92, 88, 84, *etc.*   *or*   100, 104, 108, 112, 116, 120, *etc.*

These are **consecutive numbers divisible by** 4.

1. Make a list of five consecutive numbers that are divisible by 7, counting down from 686.

2. Fill in the patterns. Notice the patterns in the *remainders*!

a. $26 \div 4 = $ _____ R _____	**b.** $78 \div 3 = $ _____ R _____	**c.** $54 \div 7 = $ _____ R _____
$27 \div 4 = $ _____ R _____	$79 \div 3 = $ _____ R _____	$55 \div 7 = $ _____ R _____
$28 \div 4 = $ _____ R _____	$80 \div 3 = $ _____ R _____	$56 \div 7 = $ _____ R _____
$29 \div 4 = $ _____ R _____	$81 \div 3 = $ _____ R _____	$57 \div 7 = $ _____ R _____
$30 \div 4 = $ _____ R _____	$82 \div 3 = $ _____ R _____	$58 \div 7 = $ _____ R _____
$31 \div 4 = $ _____ R _____	$83 \div 3 = $ _____ R _____	$59 \div 7 = $ _____ R _____
$32 \div 4 = $ _____ R _____	$84 \div 3 = $ _____ R _____	$60 \div 7 = $ _____ R _____

3. Here is a fact: 686 is evenly divisible by 7.

   a. What is the remainder if 687 is divided by 7?

   b. What is the remainder if 689 is divided by 7?

4. Here is a fact: 1,881 is evenly divisible by 11. What is the remainder if 1,886 is divided by 11?

5. If 537 leaves a remainder of 3 when divided by 6, what is the next number that is greater than 537 and is divisible by 6?

A number is divisible by 4 if the number formed from its last two digits is divisible by 4.

Example 2. To check if 5,789 is divisible by 4, just look at its last two digits, 89. Since 89 is not divisible by 4, neither is 5,789.

Why does this work? Every multiple of hundred (such as 100, 200, 300) is divisible by 4, which means that 5,700 is divisible by 4 also. Then we can simply count by fours starting at 5,700 to find numbers that are divisible by 4: 5,704; 5,708; 5,712; and so on. These correspond to this list of numbers that are divisible by four: 4, 8, 12, and so on.

6. Mark an "x" if the number is divisible by 2, 3, 4, 5, 6, or 9.

Divisible by	2	3	4	5	6	9
1,755						
298						
4,000						
3,270						

Divisible by	2	3	4	5	6	9
3,548						
277						
237						
10,999						

7. **a.** Find a number that leaves a remainder of 1 when divided by 6, and is between 90 and 100.

b. Find a number that is less than 50, and leaves a remainder of 1 when divided by 3, 4, 6, or 9.

8. Find a path through the maze <u>from the left side to the right</u>! Here are the rules:

- You may move right, left, up, or down, but not diagonally.
- Each number on your path has to be **divisible by 4**.
- Each number on your path has to be greater than the previous number on your path.

18	52	100	502	300	312	348	322
16	44	64	446	292	144	360	422
6	16	72	292	280	266	436	232
86	94	104	144	216	204	568	522
60	54	128	132	244	286	572	588
12	8	12	90	308	312	78	544
15	12	136	98	254	308	348	548
44	48	66	166	256	388	428	444

9. Find the mystery numbers! *Who am I?*

a. "I'm divisible by 8 but not by 5. I'm greater than 25 but less than 45."

b. "I am divisible by 6. I am greater than 200 but less than 220. The sum of my digits is 3."

c. "I am between 50 and 100. I am divisible by 3 and by 4. My tens digit is double my ones digit."

d. "You'll find me between 110 and 140. I don't end in a zero. And I am divisible by 12."

Prime Numbers

1. Mark an X if the number is divisible by the given numbers.

number	divisible by 1	divisible by 2	divisible by 3	divisible by 4	divisible by 5	divisible by 6	divisible by 7	divisible by 8	divisible by 9	divisible by 10
2										
3										
4										
5										
6										
7										
8										
9										
10										
11										
12										
13										
14										
15										
16										
17										
18										
19										
20										
21										
22										
23										
24										
25										
26										
27										
28										
29										
30										
31										
32										
33										
34										
35										

2. Now, find each number in this list that is only divisible by 1 and by itself. For example, 7 is one such number: it is only divisible by 1 and by 7. Such numbers are called **primes**.

Prime numbers: _____

A number is <u>prime</u> if the only way to write it as a product is 1 times the number itself.

For example, 11 is prime, because the only way to write 11 as a product is 1 × 11.
But, 12 is not a prime, because we can write it as 2 × 6. We say 12 is **composite**.
It is "composed" or "built" from other numbers by multiplication.

Example 1.
Is 450 prime or composite?

Since it ends in 0, it is divisible
by 10. Indeed, 450 = 45 × 10.
So, 450 is composite.

Example 2.
Is 88 prime or composite?

Since it is an even number, it is
divisible by 2, and we can write it as
88 = 2 × 44. So, 88 is composite.

Example 3. Is 37 prime or composite?

Check if it is divisible by 2, 3, 4, 5, 6, 7, 8, 9, or 10.

It is not divisible by 2. Because of that, it cannot be divisible
by 4, 6, 8, or 10 either, so we won't need to check those.

It is not divisible by 3 (36 is, and 37 is one more than that).
Then it cannot be divisible by 9 either.

It is not divisible by 5 since it doesn't end in 0 or 5.

It is not divisible by 7. Why? We know 35 is divisible by 7,
so 37 leaves a remainder of 2 when divided by 7.

So, 37 is prime.

3. Check if each number is prime or composite. If it is composite, write it as a multiplication.

a. 33 is prime/composite	b. 52 is prime/composite	c. 41 is prime/composite
If composite: 33 = _____ × _____	If composite: 52 = _____ × _____	If composite: 41 = _____ × _____
d. 39 is prime/composite	e. 43 is prime/composite	f. 45 is prime/composite
If composite: 39 = _____ × _____	If composite: 43 = _____ × _____	If composite: 45 = _____ × _____

Here is one more divisibility rule to help you:
A number is divisible by 3 if the sum of its digits is divisible by 3.

Example 4. Is 92 divisible by 3?
Add its digits: 9 + 2 = 11.
Since 11 is *not* divisible by 3, neither is 92.

Example 5. Is 378 divisible by 3?
Add its digits: 3 + 7 + 8 = 18.
Since 18 *is* divisible by 3, so is 378.

4. Mark an "x" if the number is divisible by 3.

number	digit sum	divisible by 3?
98		
105		
567		
59		

number	digit sum	divisible by 3?
888		
1,045		
1,338		
612		

There is no easy divisibility test for 7. (There is one, but it is not simple to use.) For numbers below 100, you can use the multiplication table of 7, which you should know by heart up to $7 \times 12 = 84$. Beyond that, add 7 to get the two other numbers that are divisible by 7: 91 and 98. You can always use long division to check if a number is divisible by 7.

5. Mark an "x" if the number is divisible by 7.

number	divisible by 7?
99	
74	
56	

number	divisible by 7?
24	
100	
84	

number	divisible by 7?
85	
63	
105	

To check if a number that is between 10 and 100 is prime or composite, it is enough to **check if it is divisible by the primes 2, 3, 5, or 7.** If it is *not* divisible by any of these, it is prime.

6. Are these numbers primes or composites? Use the divisibility rules for 2, 3, and 5 to help you.

a. 67 is prime/composite If composite: 67 = _____ × _____	**b.** 57 is prime/composite If composite: 57 = _____ × _____	**c.** 47 is prime/composite If composite: 47 = _____ × _____
d. 53 is prime/composite If composite: 53 = _____ × _____	**e.** 63 is prime/composite If composite: 63 = _____ × _____	**f.** 61 is prime/composite If composite: 61 = _____ × _____
g. 93 is prime/composite If composite: 93 = _____ × _____	**h.** 85 is prime/composite If composite: 85 = _____ × _____	**i.** 91 is prime/composite If composite: 91 = _____ × _____
j. 87 is prime/composite If composite: 87 = _____ × _____	**k.** 79 is prime/composite If composite: 79 = _____ × _____	**l.** 97 is prime/composite If composite: 97 = _____ × _____

Epilogue: Is 1 a prime number?

Up until 1899, mathematicians listed 1 as a prime number. Since then, modern mathematics has excluded 1 from the list of primes. So in today's books, the list of primes starts from 2. However, even today, some mathematicians insist 1 is a prime.

When 1 is excluded, many theorems and results of mathematics can be written in a simpler way. Fundamentally, the idea of not listing 1 as a prime is a matter of convention and convenience.

Please see also

https://www.scientificamerican.com/blog/roots-of-unity/why-isnt-1-a-prime-number/

https://en.wikipedia.org/wiki/Prime_number#Primality_of_one

Finding Factors

> **Example 1.** We can write the number 30 as a multiplication in many different ways:
>
> $30 = 10 \times 3$ and $30 = 2 \times 15$ and $30 = 5 \times 6$. There is yet one more way: $30 = 1 \times 30$.
>
> From this we learn that 10, 3, 2, 15, 5, 6, 1, and 30 are divisors or factors of 30.
>
> What about 7? Well, 30 is *not* divisible by 7, so 7 is not a factor of 30.
>
> It turns out that 1, 2, 3, 5, 6, 10, 15, and 30 are <u>ALL</u> the factors of 30. No other numbers are.

1. Find all the factors of the given numbers. Think of writing the number as a multiplication in many different ways. Don't forget the number itself times 1!

a. 6 factors:	**b.** 10 factors:
c. 12 factors:	**d.** 15 factors:
e. 20 factors:	**f.** 18 factors:

2. These students worked and found all the factors of the given numbers. But is their work correct? Be a teacher detective, and check and correct their work.

a. Aiden found all the factors of 34: $34 = 2 \times 18$ $34 = 1 \times 17$ The factors are 1, 2, 17, 18.	**b.** Olivia found all the factors of 28: $28 = 1 \times 28 \qquad 28 = 2 \times 14$ $28 = 4 \times 7$ The factors are 1, 2, 4, 7, 14, and 28.
c. Jayden found all the factors of 33: $33 = 1 \times 33$ $33 = 3 \times 13$ The factors are 1, 3, 13, 33.	**d.** Isabella found all the factors of 36: $36 = 6 \times 6$ $36 = 4 \times 9$ The factors are 4, 6, and 9.

Example 2. Find all the factors of 85.

Now, it helps to be organized. Let's check if 85 is divisible by all the numbers from 1 to 10.

- It is divisible by 1 (all numbers are): 85 = 1 × 85.

- It is not divisible by 2. Neither by 3 (its digits add up to 13). Of course it can't be divisible by 4, 6, 8, or 10 since it is not even. And it can't be divisible by 9 since it wasn't by 3.

- It *is* divisible by 5. 85 = 5 × 17. And here we can see it is also divisible by 17.

- Is it divisible by 7? No, because 84 is.

Our check is complete. So, we found <u>1, 5, 17, and 85</u>. Those are all the factors of 85.

Why do we <u>not</u> have to check if 85 is divisible by 11, 12, 13, and so on?
Because *if* 85 was 11 times some number, it would be 11 times some *smaller* number than 11. We went through all the smaller numbers already and didn't find that any of them times 11 was 85.

3. Find all the factors of the given numbers.

a. 46 Check 1 2 3 4 5 6 7 8 9 10 factors: _____	**b.** 68 Check 1 2 3 4 5 6 7 8 9 10 factors: _____
c. 99 Check 1 2 3 4 5 6 7 8 9 10 factors: _____	**d.** 72 Check 1 2 3 4 5 6 7 8 9 10 factors: _____
e. 73 Check 1 2 3 4 5 6 7 8 9 10 factors: _____	**f.** 80 Check 1 2 3 4 5 6 7 8 9 10 factors: _____
g. 95 Check 1 2 3 4 5 6 7 8 9 10 factors: _____	**h.** 64 Check 1 2 3 4 5 6 7 8 9 10 factors: _____

Primes and Finding Factors

Let's now review how to find all the factors of a given number.

Example 1. Find all the factors of 92.

Now, it helps to be organized. Let's check if 92 is divisible by all the numbers from 1 to 10, and keep track of the factors we find.

- It *is* divisible by 1 (all numbers are): $92 = 1 \times 92$. So, **1** and **92** are factors of 92.

- It *is* divisible by 2: $92 = 2 \times 46$. So, here we find **2** and **46** both are factors of 92.

- It is *not* divisible by 3 (the digit sum is 11). It cannot be divisible by 6 or 9 since it was not by 3.

- It *is* divisible by 4, because we can skip-count from it and reach 100, which clearly is divisible by 4. We write $92 = 4 \times 23$. So, **4** and **23** are factors of 92.

- It is *not* divisible by 5 or by 10 as it does not end in 0 nor 5.

- Is it divisible by 7? No, because 84, 91, and 98 are.

- By 8? Skip-count from 80 by eights: 80, 88, and 96 are divisible by 8. So, 92 is not.

Our check is complete. So, we found 1, 2, 4, 23, 46, and 92. Those are all the factors of 92.

Why do we not have to check if 92 is divisible by 11, 12, 13, and so on?
If 92 was 11 times a number, it would be 11 times some *smaller* number than 11. We went through all the smaller numbers already and did not find that any of them times 11 would equal 92.

1. Find all the factors of the given numbers. Use the checklist; keep track of *all* the factors you find.

a. 26 Check 1 2 3 4 5 6 7 8 9 10 factors: _____	**b.** 38 Check 1 2 3 4 5 6 7 8 9 10 factors: _____
c. 88 Check 1 2 3 4 5 6 7 8 9 10 factors: _____	**d.** 47 Check 1 2 3 4 5 6 7 8 9 10 factors: _____
e. 71 Check 1 2 3 4 5 6 7 8 9 10 factors: _____	**f.** 86 Check 1 2 3 4 5 6 7 8 9 10 factors: _____

In this table, we show numbers from 2 through 10 and what numbers they are divisible by.

Some rows are highlighted, because those numbers have only two factors: 1 and the number itself.

Those numbers are called **prime numbers**, or just **primes**.

Prime numbers less than 10 are 2, 3, 5, and 7.

What is the next prime after 10?

Number	Divisible by:									
	1	2	3	4	5	6	7	8	9	10
2	x	x								
3	x		x							
4	x	x		x						
5	x				x					
6	x	x	x			x				
7	x						x			
8	x	x		x				x		
9	x		x						x	
10	x	x			x					x

What about number 1? Number 1 is *not* **a prime.** Please see the note at the end of the lesson to learn more.

2. For each number in the table, find all its factors. Note the numbers that only have two factors: one and the number itself. Those are primes.

Number	Factors
11	
12	
13	
14	
15	
16	
17	
18	
19	
20	

3. Write a list of primes between 1 and 20: _____

   Here is a list of primes between 20 and 50: 23, 29, 31, 37, 41, 43, 47.

4. (*Optional*) Visit **https://www.mathmammoth.com/practice/sieve-of-eratosthenes**
   for an interactive tool that will find primes using a sieving process.

Why are primes so special? Because it turns out that *every* whole number can be written as a multiplication, using primes only! This is called the prime factorization of a number.

For example, $730 = 2 \times 5 \times 73$. Each of the factors, 2, 5, and 73 are primes.

Or, $2,904 = 2 \times 2 \times 2 \times 3 \times 11 \times 11$.

And this factorization is unique for each number; there is no other way to do it.

This fact has important applications in computer security and cryptography.

5. **a.** Find a prime between 50 and 60.

b. Find a prime between 80 and 90.

a. Find a prime between 110 and 120.

Puzzle Corner

b. Number 24 has eight factors: 1, 2, 3, 4, 6, 8, 12, and 24.
Find a number that has even more factors and is less than 40.

c. Find a number that is divisible by 3 and by 5 and has exactly eight factors.

Is 1 a prime number?

Up until 1899, mathematicians listed 1 as a prime number. Since then, modern mathematics has excluded 1 from the list of primes. So in today's books, the list of primes starts from 2. However, even today, some mathematicians insist 1 is a prime.

When 1 is excluded, many theorems and results of mathematics can be written in a simpler way, but fundamentally, the idea of not listing 1 as a prime is a matter of convention and convenience.

Please see also:

https://www.scientificamerican.com/blog/roots-of-unity/why-isnt-1-a-prime-number/

https://en.wikipedia.org/wiki/Prime_number#Primality_of_one

Prime Factorization

Prime numbers have only two divisors: 1 and the number itself. If a number is not prime, it is a **composite number**. In the last lesson, we found that the primes less than 30 are **2, 3, 5, 7, 11, 13, 17, 19, 23,** and **29**.

When you write a number as a product, you are **factoring** the number. For example, we can write 96 as $96 = 3 \times 32$, and we have factored 96. Another way to factor 96 is $96 = 6 \times 4 \times 4$. But now we will look at a very special way to factor a number: its **prime factorization**: a way to factor the number that will *only* use primes!

A factor tree is a handy way to factor composite numbers to their prime factors. The factor tree starts at the root and grows "upside down!"

We write 24 on top. First, 24 is written as 4×6. However, 4 and 6 are not primes, so we continue. Four is factored into 2×2 and six is factored into 2×3.

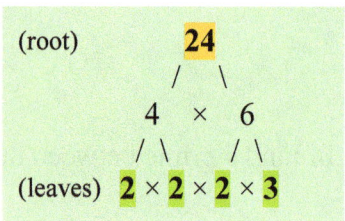

```
(root)          24
               /  \
              4  ×  6
             / \   / \
(leaves)   2 × 2 × 2 × 3
```

We cannot factor 2 or 3 any further because they are prime numbers. Once you get to primes in your "tree," they are the "leaves" and you stop factoring in that "branch." So the **prime factorization of 24 is: $24 = 2 \times 2 \times 2 \times 3$.**

Examples of factoring some composite numbers:

```
    30
   /  \
  5  ×  6
      / \
     2 × 3
```
5 is a prime number or a "leaf." Once you're done, "pick the leaves"—you can circle them to see them better! So, $30 = 2 \times 3 \times 5$.

```
    66               66
   /  \             /  \
 11  ×  6   OR    2  ×  33
      / \              / \
     2 × 3           11 × 3
```
You can start the factoring process any way you want. The end result is the same: $66 = 2 \times 3 \times 11$.

```
    21
   /  \
  3  ×  7
```
Both 3 and 7 are prime numbers, so we cannot factor them any further. $21 = 3 \times 7$.

```
    89
   /  \
  1  ×  89
```
The only way to write 89 as a product of primes is 1×89. This means it is prime.

```
       72
      /  \
    12  ×  6
   / \   / \
  3 × 4 × 2 × 3
     / \
    2 × 2
```
Seventy-two has lots of factors so the factoring takes many steps.

We get $72 = 2 \times 2 \times 2 \times 3 \times 3$.

We could have also started with $72 = 2 \times 36$ or $72 = 4 \times 18$.

```
    57
   /  \
```
How do you get started?

- Check if 57 is in any of the times tables.

- Use divisibility tests to check if it is divisible by 2, 3, 4, 5, etc.

1. Factor the following composite numbers to their prime factors.

a. 18
/ \\
2 × 9
  / \\
  3 × 3
 	
18 = 2 × 3 × 3
	b. 6
	/\\
		c. 14
		/\\

d. 8 /\\	**e.** 12 /\\	**f.** 20 /\\
g. 16 /\\	**h.** 24 /\\	**i.** 27 /\\
j. 25 /\\	**k.** 33 /\\	**l.** 15 /\\

2. Find the prime factorization of the numbers. If the number is prime, write it as 1 times the number.

a. 42 / \	**b.** 56 / \	**c.** 68 / \
d. 75 / \	**e.** 47 / \	**f.** 99 / \
g. 72 / \	**h.** 80 / \	**i.** 97 / \
j. 85 / \	**k.** 66 / \	**l.** 82 / \

Prime numbers are like building blocks of all numbers. They are the first and foremost, and the rest of the numbers—the composite numbers—are "built" from them. "Building numbers" is like factoring backwards. We start with the building blocks (the primes) and see what we get:

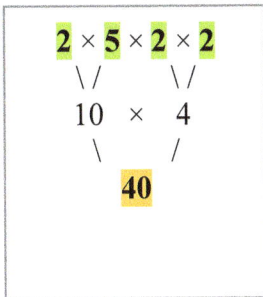

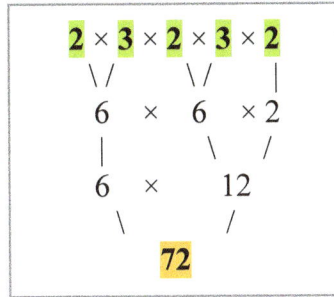

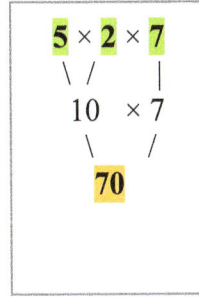

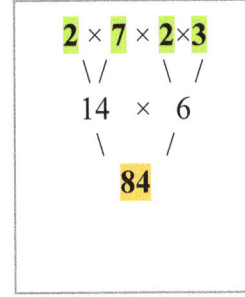

```
2 × 5 × 2 × 2          2 × 3 × 2 × 3 × 2          5 × 2 × 7          2 × 7 × 2×3
 \ /     \ /            \ /     \ /   |           \ /     |          \ /     \ /
 10  ×    4             6   ×   6  × 2            10   × 7           14   ×   6
   \     /              |        \    /             \    /            \      /
     40                 6   ×       12                70                 84
                         \          /
                            72
```

By using the process above (building numbers starting from primes) you can build *any* whole number there is! Can you believe that?

Stated in another way: **ALL numbers** can be factored so that the factors are prime numbers. That is sort of amazing! This fact is known as the *fundamental theorem of arithmetic*. Indeed, it is fundamental.

So, no matter what the number is—992 or 83,283 or 150,282— it can be written as a product of primes.

```
        992
        / \
     4  ×  248
    / \    /  \
   2 × 2 × 4 × 62
         / \  / \
        2 × 2 × 2 × 31
```

The number 992 = 2 × 2 × 2 × 2 × 2 × 31 (see the factorization on the right). For 83,283 we get 3 × 17 × 23 × 71, and 150,282 = 2 × 3 × 3 × 3 × 11 × 11 × 23.

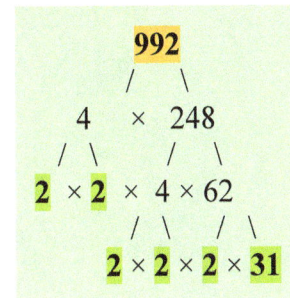

To find these factorizations, you need to test-divide the numbers by various primes, which is a bit tedious. That is why people use computers to help with factorization of numbers.

3. Build composite numbers from primes.

a. 2 × 5 × 11
  \ / |
	b. 3 × 2 × 2 × 2
	   \ / \ /
		c. 2 × 3 × 7
		  \ / |

d. 11 × 3 × 2
  | \ /
	e. 3 × 3 × 2 × 5
	   \ / \ /
		f. 2 × 3 × 17

4. Build more composite numbers from primes.

a. $2 \times 5 \times 13$	b. $7 \times 13 \times 2 \times 11$	c. $19 \times 3 \times 5 \times 2$

5. Try it on your own! Pick 3-6 primes as you wish (you can use the same prime several times), and see what number is built from them.

Puzzle Corner Ready for a challenge? Use your knowledge of divisibility tests and the calculator, and find the prime factorization of these numbers:

a. 2,145	b. 3,680	c. 10,164

The Sieve of Eratosthenes and Prime Factorization

Remember? A number is a **prime** if it has no other factors besides 1 and itself.

For example, 13 is a prime, since the only way to write it as a multiplication is 1 · 13. In other words, 1 and 13 are its only factors.

And, 15 is not a prime, since we can write it as 3 · 5. In other words, 15 has other factors besides 1 and 15, namely 3 and 5.

To find all the prime numbers less than 100 we can use the *sieve of Eratosthenes*.

Here is an online interactive version: https://www.mathmammoth.com/practice/sieve-of-eratosthenes

1. Cross out 1, as it is not considered a prime.
2. Cross out all the even numbers except 2.
3. Cross out all the multiples of 3 except 3.
4. You do not have to check multiples of 4. Why?
5. Cross out all the multiples of 5 except 5.
6. You do not have to check multiples of 6. Why?
7. Cross out all the multiples of 7 except 7.
8. You do not have to check multiples of 8 or 9 or 10.
9. The numbers left are primes.

1	2	3	4	5	6	7	8	9	10
11	12	13	14	15	16	17	18	19	20
21	22	23	24	25	26	27	28	29	30
31	32	33	34	35	36	37	38	39	40
41	42	43	44	45	46	47	48	49	50
51	52	53	54	55	56	57	58	59	60
61	62	63	64	65	66	67	68	69	70
71	72	73	74	75	76	77	78	79	80
81	82	83	84	85	86	87	88	89	90
91	92	93	94	95	96	97	98	99	100

List the **primes between 0 and 100** below:

2, 3, 5, 7, _____

Why do you not have to check numbers that are bigger than 10? Let's think about multiples of 11. The following multiples of 11 have already been crossed out: 2 · 11, 3 · 11, 4 · 11, 5 · 11, 6 · 11, 7 · 11, 8 · 11 and 9 · 11. The multiples of 11 that have not been crossed out are 10 · 11 and onward... but they are not on our chart! Similarly, the multiples of 13 that are less than 100 are 2 · 13, 3 · 13, ..., 7 · 13, and all of those have already been crossed out when you crossed out multiples of 2, 3, 5 and 7.

1. You learned this in grades 4 and 5... find all the factors of the given numbers. Use the checklist to help you keep track of which factors you have tested.

a. 54 Check 1 2 3 4 5 6 7 8 9 10 factors: _____	**b.** 60 Check 1 2 3 4 5 6 7 8 9 10 factors: _____
c. 84 Check 1 2 3 4 5 6 7 8 9 10 factors: _____	**d.** 97 Check 1 2 3 4 5 6 7 8 9 10 factors: _____

A number is...

divisible by 2 if it ends in 0, 2, 4, 6, or 8.

divisible by 5 if it ends in 0 or 5.

divisible by 10 if it ends in 0.

divisible by 100 if it ends in "00".

A number is...

divisible by 3 if the sum of its digits is divisible by 3.

divisible by 4 if the number formed from its last two digits is divisible by 4.

divisible by 6 if it is divisible by both 2 and 3.

divisible by 9 if the sum of its digits is divisible by 9.

Use the various divisibility tests when building a factor tree for a composite number.

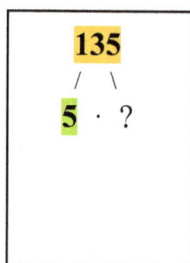 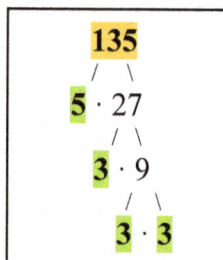 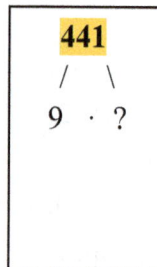 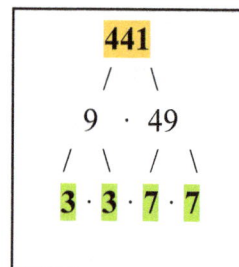

We start out by noticing that 135 is **divisible by 5.** From long division, we get $135 = 5 \cdot 27$. The final factorization is $135 = 3 \cdot 3 \cdot 3 \cdot 5$ or $3^3 \cdot 5$.

Adding the digits of 441, we get 9, so it is **divisible by 9.** We divide to get $441 = 9 \cdot 49$. The end result is $441 = 3 \cdot 3 \cdot 7 \cdot 7$ or $3^2 \cdot 7^2$.

2. Find the prime factorization of these composite numbers. Use a notebook for long divisions. Give each factorization below the factor tree.

a. 124	b. 260	c. 96
2 · ___ / \ 2	10 · ___ / \ / \	3 · ___ / \
124 =	**260 =**	**96 =**
d. 90	e. 165	f. 95
90 =	**165 =**	**95 =**

3. Mark an "x" if the number is divisible by 2, 3, 4, 5, 6, or 9.

Divisible by	2	3	4	5	6	9
128						
765						

Divisible by	2	3	4	5	6	9
209						
6,042						

4. Find the prime factorization of the numbers. Use a notebook for long divisions. Give each factorization below the factor tree.

Note: in (a), the last two digits of 912 are "12" so it is **divisible by 4**.

a. 912
/ \
4 · _____

b. 528

912 =

528 =

c. 76

d. 126

e. 272

76 =

126 =

272 =

5. Mia and Alex found the prime factorization of 164 and 168, and were completely surprised that they got the same factorization for both!

Investigate the situation. Is there something fishy going on somewhere?

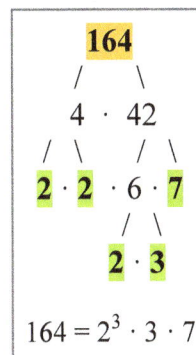

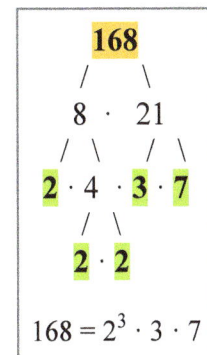

```
      164                    168
     /   \                  /   \
    4  ·  42               8  ·  21
   / \   / \              / \   / \
  2·2 · 6·7              2·4 · 3·7
         / \                / \
        2·3               2·2
164 = 2³ · 3 · 7      168 = 2³ · 3 · 7
```

6. Find all the primes between 100 and 110. How? You need to check, for each number, whether it is divisible by 2, 3, 4, 5, 6, 7, 8, 9, or 10.

7. Find the prime factorization of these composite numbers.

a. 196	b. 380	c. 336
196 =	380 =	336 =
d. 306	e. 116	f. 720
306 =	116 =	720 =
g. 675	h. 990	i. 945
675 =	990 =	945 =

Puzzle Corner   Find all the primes between 0 and 200. Use the sieve of Eratosthenes again (you need to make a grid in your notebook).

This time, you need to cross out 1, and then every even number except 2, every multiple of 3 except 3, every multiple of 5 except 5, every multiple of 7 except 7, every multiple of 11 except 11 and every multiple of 13 except 13.

The Greatest Common Factor (GCF)

Let's take two whole numbers. We can then list all the <u>factors</u> of each number, and then find the factors that are <u>common</u> in both lists. Lastly, we can choose the <u>greatest</u> or largest among those "common factors." That is the **greatest common factor** of the two numbers. The term itself really tells you what it means!

Example 1. Find the greatest common factor of 18 and 30.

<u>The factors of 18:</u> 1, 2, 3, 6, 9 and 18.
<u>The factors of 30:</u> 1, 2, 3, 5, 6, 10, 15 and 30.

Their <u>common</u> factors are 1, 2, 3 and 6. The <u>greatest</u> common factor is 6.

Here is a **method to find all the factors of a given number**.

Example 2. Find the factors (divisors) of 36.

We check if 36 is divisible by 1, 2, 3, 4 and so on. Each time we find a divisor, we write down *two* factors.

- 36 is divisible by 1. We write $36 = 1 \cdot 36$, and that equation gives us two factors of 36: both the smallest (**1**) and the largest (**36**).

- 36 is also divisible by 2. We write $36 = 2 \cdot 18$, and that equation gives us two more factors of 36: the second smallest (**2**) and the second largest (**18**).

- Next, 36 is divisible by 3. We write $36 = 3 \cdot 12$, and now we have found the third smallest factor (**3**) and the third largest factor (**12**).

- Next, 36 is divisible by 4. We write $36 = 4 \cdot 9$, and we have found the fourth smallest factor (**4**) and the fourth largest factor (**9**).

- Finally, 36 is divisible by 6. We write $36 = 6 \cdot 6$, and we have found the fifth smallest factor (**6**) which is also the fifth largest factor.

We know that we are done because the list of factors from the "small" end (1, 2, 3, 4, 6) has met the list of factors from the "large" end (36, 18, 12, 9, 6).

Therefore, all of the factors of 36 are: 1, 2, 3, 4, 6, 9, 12, 18 and 36.

1. List all of the factors of the given numbers.

a. 48	**b.** 60
c. 42	**d.** 99

2. Find the greatest common factor of the given numbers. Your work above will help!

a. 48 and 60	**b.** 42 and 48	**c.** 42 and 60	**d.** 99 and 60

3. List all of the factors of the given numbers.

a. 44	b. 66
c. 28	d. 56
e. 100	f. 45

4. Find the greatest common factor of the given numbers. Your work above will help!

a. 44 and 66	b. 100 and 28	c. 45 and 100	d. 45 and 66
e. 28 and 44	f. 56 and 28	g. 56 and 100	h. 45 and 28

Example 3. What is the greatest common factor useful for?

It can be used to simplify fractions. For example, let's say you know that the GCF of 66 and 84 is 6. Then, to simplify the fraction 66/84 to lowest terms, you divide both the numerator and the denominator by 6. →

$$\overset{\div 6}{\frac{66}{84}} = \frac{11}{14} \quad \div 6$$

However, it is *not* necessary to use the GCF when simplifying fractions. You can always simplify in several steps. See the example at the right. →

Or, you can *simplify by factoring*, like we did in the previous lesson:

$$\frac{66}{84} = \frac{6 \cdot 11}{7 \cdot 6 \cdot 2} = \frac{11}{14}$$

In fact, these other methods might be quicker than using the GCF.

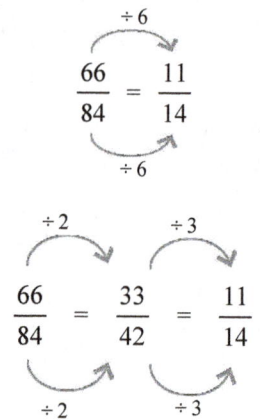

$$\overset{\div 2 \quad\quad \div 3}{\frac{66}{84}} = \frac{33}{42} = \frac{11}{14} \quad \div 2 \quad\quad \div 3$$

---

5. Simplify these fractions, if possible. Your work in the previous exercises can help!

a. $\dfrac{48}{66}$ b. $\dfrac{42}{44}$ c. $\dfrac{42}{48}$ d. $\dfrac{99}{60}$

e. $\dfrac{48}{100}$ f. $\dfrac{100}{99}$ g. $\dfrac{56}{28}$ h. $\dfrac{44}{99}$

Using prime factorization to find the greatest common factor (optional)

Another, more efficient way to find the GCF of two or more numbers is to use the prime factorizations of the numbers to find *all* of the common prime factors. The product of those common prime factors forms the GCF.

Example 4. Find the GCF of 48 and 84.

The prime factorizations are: $48 = 2 \cdot 2 \cdot 2 \cdot 2 \cdot 3$ and $84 = 2 \cdot 2 \cdot 3 \cdot 7$.

We see that the common prime factors are 2 and 2 and 3. So, the GCF is $2 \cdot 2 \cdot 3 = 12$.

Example 5. Find the GCF of 75, 105 and 150.

The prime factorizations are: $75 = 3 \cdot 5 \cdot 5$, $105 = 3 \cdot 5 \cdot 7$ and $150 = 2 \cdot 3 \cdot 5 \cdot 5$.

The common prime factors for all of them are 3 and 5. So, the GCF of these three numbers is $3 \cdot 5 = 15$.

6. Find the greatest common factor of the numbers.

a. 120 and 66	**b.** 36 and 136
c. 98 and 76	**d.** 132 and 72
e. 45 and 76	**f.** 64 and 120

7. Find the greatest common factor of the given numbers.

a. 75, 25 and 90	**b.** 54, 36 and 40
c. 18, 24 and 36	**d.** 72, 60 and 48

Find the greatest common factor of 187 and 264.

Puzzle Corner

Factoring Sums

1. You have seen these before! Write two different expressions for the total area, thinking of: (1) the area of the big rectangle as a whole and (2) the sum of the areas of the two small rectangles.

a. 3(_____ + _____) and

3 · _____ + 3 · _____ =

b. _____ (_____ + _____) and

_____ · _____ + _____ · _____ =

c. _____ (_____ + _____) and

_____ · _____ + _____ · _____ =

d. _____ (_____ + _____) and

_____ · _____ + _____ · _____ =

2. List the length and width of all the possible rectangles with an area of 30 cm^2 and where the sides are whole numbers (in centimeters). The first one would be 1 cm × 30 cm, and the second would be 2 cm × 15 cm.

3. List the length and width of all of the possible rectangles that have an area of 40 m^2 and where the sides are whole numbers (in meters). The first one would be 1 m × 40 m.

4. Build a big rectangle out of two smaller ones like in exercise #1. Choose one rectangle from exercise #2 and one from #3 that each have one side that is the same length. In the grid at the right, sketch the two rectangles that you chose side by side, touching so that they share the side that is the same length. Your sketch should look like the figures in exercise #1.

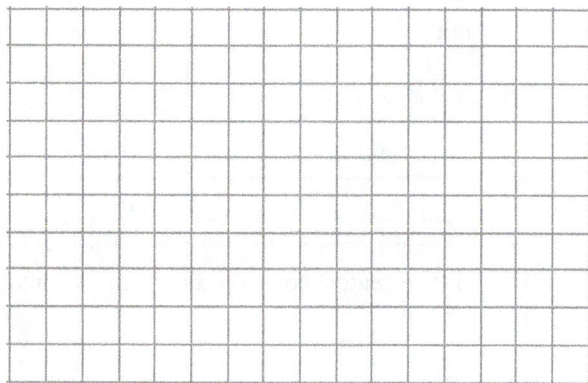

5. Can you find another answer to exercise 4?

Another usage for the GCF is to factor expressions. **Factoring an expression** means writing it as a product (a multiplication).

Example 1. We can easily write the sum $54 + 27$ as multiplication, once we notice that both 54 and 27 have the factor 9. So, $54 + 27$ is $9 \cdot 6 + 9 \cdot 3$. Now, using the distributive property "backwards," we write

$$9 \cdot 6 + 9 \cdot 3 = 9(6 + 3)$$

We have now <u>factored</u> the original sum. This means we have <u>written it as a multiplication</u>. This time we have two factors: the first factor is 9, and the second factor is actually a *sum*: the sum $6 + 3$.

Example 2. Write the sum $92 + 56$ as a multiplication, using the greatest common factor of 92 and 56.

The GCF of 92 and 56 is 4. So, we write $92 + 56 = 4 \cdot 23 + 4 \cdot 14 = 4(23 + 14)$.

Example 3. You have also factored expressions such as $72x + 16$ using the distributive property "backwards." Notice, the GCF of 72 and 16 is 8. We can write $72x + 16$ as $8(9x + 2)$.

6. First find the GCF of the numbers. Then factor the expressions using the GCF.

a. GCF of 18 and 12 is `6`

$18 + 12 =$ `6` $\cdot 3 +$ `6` $\cdot 2 =$ `6` (____ + ____)

b. GCF of 6 and 10 is `____`

$6 + 10 =$ `____` $\cdot$ ____ + `____` $\cdot$ ____ = `____` (____ + ____)

c. GCF of 22 and 11 is `____`

$22 + 11 =$ `____` $\cdot$ ____ + `____` $\cdot$ ____ = `____` (____ + ____)

d. GCF of 15 and 21 is _____

$15 + 21 =$ ____ $\cdot$ ____ + ____ $\cdot$ ____ = ____ (____ + ____)

e. GCF of 25 and 35 is _____

$25 + 35 =$ ____ (____ + ____)

f. GCF of 72 and 86 is _____

$72 + 86 =$ ____ (____ + ____)

g. GCF of 96 and 40 is _____

$96 + 40 =$ ____ (____ + ____)

h. GCF of 39 and 81 is _____

$39 + 81 =$ ____ (____ + ____)

7. **a.** Express the sum 32 + 40 as a product (multiplication).

   b. Draw two rectangles, side by side, to represent the product you wrote.

8. Draw two rectangles, side by side, to represent the sum 30 + 25.

9. Draw three rectangles, side by side, to represent the sum 42 + 24 + 30.

10. You need to build a rectangular animal pen that has an area of 45 m^2.
    If the lengths of the sides need to be in whole meters, what are the options?

11. **a.** List a pair of numbers whose greatest common factor is 1.

    b. List two more pairs of numbers whose greatest common factor is 1.

12. Write these sums as a product (multiplication) of their GCF and another sum.

a. The GCF of 15 and 5 is _____ $15x + 5 =$ ____ (____ + ____)	**b.** The GCF of 18 and 30 is _____ $18x + 30 =$ ____ (____ + ____)
c. The GCF of 72 and 54 is _____ $72a + 54b =$ ____ (____ + ____)	**d.** The GCF of 100 and 90 is _____ $100y + 90x =$ ____ (____ + ____)

a. List two numbers whose greatest common factor is 13.

Puzzle Corner

b. List two numbers whose greatest common factor is 51.

The Least Common Multiple (LCM)

A **multiple** of a whole number n is any of the numbers n, $2n$, $3n$, $4n$, $5n$ and so on. In other words, a whole number times the number n is a multiple of n.

Example 1. The multiples of 9 are 9, 18, 27, 36, 45, 54, 63 and so on.

When we have two or more whole numbers, we can find their **least common multiple**.

As in the case of the greatest common factor, the term "least common multiple" itself tells us what it is! Read it again: least common multiple. All we need to do (in principle) is to find the multiples of the numbers, then find the common multiples, and lastly choose the one that is the least, or the smallest.

Example 2. Find the least common multiple of 5 and 8.
- The multiples of 5 are: 5, 10, 15, 20, 25, 30, 35, <u>40</u>, 45, 50, ... <u>80</u>,...
- The multiples of 8 are: 8, 16, 24, 32, <u>40</u>, 48, 56, 64, 72, <u>80</u>, ...

Among these multiples we find the common multiples 40 and 80. There are others as well, such as 120, 160 and so on, but 40 is the least (smallest) common multiple of 5 and 8.

Also, 40 is $5 \cdot 8$. Note: $a \cdot b$ is *always* a common multiple of both a and b, but it is not always the *least* common multiple.

Example 3. Find the least common multiple of 4 and 6.
- The multiples of 4 are: 4, 8, <u>12</u>, 16, 20, <u>24</u>, 28, 32, <u>36</u>, 40, ...
- The multiples of 6 are: 6, <u>12</u>, 18, <u>24</u>, 30, <u>36</u>, 42, 48, 54, 60, ...

Among these, we find the common multiples 12, 24 and 36. The *least* common multiple (LCM) is 12.

Note that the LCM of 4 and 6 is *not* $4 \cdot 6$.

1. Find the LCM of these numbers.

a. 2 and 6	**b.** 6 and 9
c. 14 and 8	**d.** 3 and 8
e. 7 and 10	**f.** 10 and 15

2. **a.** List four multiples of 6 that are less than 100.

   b. What is the biggest multiple of 4 that is less than 100?

   c. What is the smallest multiple of 250 that is more than 1,000?

3. A bus running Route A leaves the main bus terminal at every 15 minutes, and a bus running Route B at every 12 minutes. If there was a bus for both routes leaving at 3:30 PM, when is the next time that there is a bus for both routes leaving at the same time?

4. Boxes that are 20 cm tall are being stacked, as well as boxes that are 45 cm tall. What is the least height in centimeters at which both stacks are the same height?

5. Go back to exercise #1. Is the LCM of the two numbers also their product?

a. Is the LCM of 2 and 6 equal to 2 · 6?	**b.** Is the LCM of 6 and 9 equal to 6 · 9?
c. Is the LCM of 14 and 8 equal to 14 · 8?	**d.** Is the LCM of 3 and 8 equal to 3 · 8?
e. Is the LCM of 7 and 10 equal to 7 · 10?	**f.** Is the LCM of 10 and 15 equal to 10 · 15?

You may be wondering why sometimes the LCM is the product of the two numbers, and other times it is not. The key is:

If the numbers do not have *any common factors* (except 1), then their LCM is their product.

Example 4. The numbers 8 and 10 have a common factor 2. Therefore, their LCM will not be 80. (Can you find what it is?)

6. Find the LCM of these numbers.

a. 3 and 4	**b.** 9 and 7
c. 10 and 5	**d.** 4 and 7
e. 2 and 10	**f.** 4 and 10

7. **a.** Draw a line from each number to the correct box.

   b. Which number is a "black sheep" (neither a factor nor a multiple of 24)?

   c. Which number is BOTH a factor and a multiple of 24?

240   8   48   4   96   24   1   2

a factor of 24	a multiple of 24

120   3   30   72   144   6   12

Example 5. Find the least common multiple of 4, 6, and 5.

To find the LCM of three numbers, you could make three lists of their multiples, but a quicker way is to *first* find the LCM of two of the numbers, and then use that.

The LCM of 4 and 6 is 12. Now we will find the LCM of 12 and 5, and that will be the LCM of 4, 6, and 5.

Since 12 and 5 don't have common factors, their LCM is 12 · 5 = 60, and that is also the LCM of 4, 6, and 5.

8. Find the LCM of three numbers.

a. 3 and 8 and 6	**b.** 2 and 6 and 10
c. 3 and 5 and 2	**d.** 4 and 7 and 8

9. Anne is a hair stylist. Among her customers, Mrs. Goodwill comes for a haircut every 8 weeks, Ms. Sidney every 6 weeks, and Ms. Locksmith every 4 weeks. If all of them came for a haircut on a certain Monday, in how many weeks will they again have a haircut on the same day?

Remember? Before adding or subtracting *unlike* fractions, we need to convert them into equivalent fractions that have a **common denominator**.

Example 6 (on the right). The denominators 8 and 10 need to "go into" the common denominator. In other words, the common denominator must be **a multiple of both** 8 and 10. Naturally, the least common multiple is what we are looking for! The LCM of 8 and 10 is 40.

You *could* use any common multiple as the common denominator, (such as 80), but the LCM is the least (smallest) common denominator.

$$\frac{5}{8} + \frac{1}{10}$$
$$\downarrow \qquad \downarrow$$
$$\frac{25}{40} + \frac{4}{40} = \frac{29}{40}$$

10. Add or subtract the fractions. Give your answer in lowest terms.

a. $\dfrac{1}{9} + \dfrac{1}{8}$	**b.** $\dfrac{1}{12} + \dfrac{7}{8}$
c. $\dfrac{3}{7} - \dfrac{3}{10}$	**d.** $\dfrac{8}{9} - \dfrac{1}{6}$

(*This section is optional.*) There is a **quicker and more efficient way for finding the least common multiple of numbers**, where we do not have to make lists of multiples. It is based on prime factorization.

1. Write the prime factorization of the numbers.
2. The LCM is formed by taking ALL the factors from the numbers, without repeating any common factor.

Example 7. Find the LCM of 45 and 25.

1. The prime factorizations are $45 = 3 \cdot 3 \cdot 5$ and $25 = 5 \cdot 5$.

2. Form a number that is "all inclusive" or that includes all the factors from both numbers. It is $3 \cdot 3 \cdot 5 \cdot 5$, which equals 225.

Notice that $3 \cdot 3 \cdot 5 \cdot 5$ includes *both* $3 \cdot 3 \cdot 5$ and $5 \cdot 5$ but has no other factors beyond those.

Example 8. Find the LCM of 24 and 40.

1. The prime factorizations are $24 = 2 \cdot 2 \cdot 2 \cdot 3$ and $40 = 2 \cdot 2 \cdot 2 \cdot 5$.

2. The number that includes all the factors from both numbers is $2 \cdot 2 \cdot 2 \cdot 3 \cdot 5 = 120$.

11. Find the least common multiple of the numbers using any method.

a. 40 and 15	**b.** 20 and 24
c. 20 and 16	**d.** 50 and 120

12. Convert the fractions so they have the same denominator, and then compare them. Your work above can help!

a. $\dfrac{29}{40}$ and $\dfrac{11}{15}$ $\downarrow \quad\quad \downarrow$ $\square$	**b.** $\dfrac{11}{20}$ and $\dfrac{13}{24}$ $\downarrow \quad\quad \downarrow$ $\square$	**c.** $\dfrac{7}{20}$ and $\dfrac{5}{16}$ $\downarrow \quad\quad \downarrow$ $\square$	**d.** $\dfrac{39}{50}$ and $\dfrac{94}{120}$ $\downarrow \quad\quad \downarrow$ $\square$

Puzzle Corner If the first day of the year is Tuesday, what day of the week is day number 236?

Review

1. Factor the following composite numbers into their prime factors.

a. 81 /\	**b.** 26 /\	**c.** 65 /\
d. 96 /\	**e.** 124 /\	**f.** 450 /\

2. Simplify.

a. $\dfrac{28}{84} = \dfrac{4 \cdot 7}{21 \cdot 4} =$	**b.** $\dfrac{75}{160} =$
c. $\dfrac{222}{36} =$	**d.** $\dfrac{48}{120} =$

3. Find the least common multiple of these pairs of numbers.

a. 3 and 7	**b.** 10 and 8
c. 11 and 6	**d.** 6 and 8

4. Find the greatest common factor of the given number pairs.

a. 24 and 64	**b.** 100 and 75
c. 80 and 96	**d.** 78 and 96

5. Fill in with the words "multiple(s)" or "factor(s)."

a.
- 25, 50, 75, 100, 125 and 150 are _____ of 25.
- 1, 2, 5, 10, 25 and 50 are _____ of 50.
- Each number has an infinite number of _____.
- Each number has a greatest _____.
- If a number x divides into another number y, we say x is a _____ of y.

b. List five different multiples of 15 that are less than 200 but more than 60.

c. Find five numbers that are multiples of both 4 and 7.

What is the LCM of 4 and 7?

6. First, find the GCF of the numbers. Then factor the expressions using the GCF.

a. GCF of 12 and 21 is _____ $12 + 21 = \underline{\quad} \cdot \underline{\quad} + \underline{\quad} \cdot \underline{\quad} = \underline{\quad}(\underline{\quad} + \underline{\quad})$
b. GCF of 45 and 70 is _____ $45 + 70 = \underline{\quad}(\underline{\quad} + \underline{\quad})$

7. Draw two rectangles, side by side, to represent the sum 42 + 30.

Answer Key

Divisibility, pp. 7-10

1. a. 7; yes   b. 6 R4; no   c. 3 R2; no   d. 12; yes

2. a. 24 R2, no   b. 86 R1; no   c. 418 R2; no

3. Here is a multiplication fact: $8 \times 9 = 72$. So, 8 is a <u>factor</u> of 72, and so is 9.
   Also, 72 is a <u>multiple</u> of 8, and also 72 is a <u>multiple</u> of 9. And, 72 is <u>divisible</u> by 8 and also by 9.

4.

a. Is 5 a factor of 55?
   Yes, because $5 \times 11 = 55$.
	b. Is 8 a divisor of 45?
	   No, because $45 \div 8 = 5$ R5.

c. Is 36 a multiple of 6?
   Yes, because $6 \times 6 = 36$.
	d. Is 34 a multiple of 7?
	   No, because $34 \div 7 = 4$ R6.

e. Is 7 a factor of 46?
   No, because $46 \div 7 = 6$ R4.
   (It is not an even division.)
	f. Is 63 a multiple of 9?
	   Yes, because $7 \times 9 = 63$.

5. a. 0, 11, 22, 33, 44, 55, 66, 77, 88, 99, 110, 121, 132, 143, 154
   b. 0, 111, 222, 333, 444, 555, 666, 777, 888, 999, 1,110, 1,221, 1,332, 1,443, 1,554, 1,665

6.

number	divisible by 2	divisible by 5
750	x	x
751		
752	x	
753		
754	x	

number	divisible by 2	divisible by 5
755		x
756	x	
757		
758	x	
759		

number	divisible by 2	divisible by 5
760	x	x
761		
762	x	
763		
764	x	

number	divisible by 2	divisible by 5
765		x
766	x	
767		
768	x	
769		

7.

number	divisible by 2	divisible by 5	divisible by 10
860	x	x	x
861			
862	x		
863			
864	x		

number	divisible by 2	divisible by 5	divisible by 10
865		x	
866	x		
867			
868	x		
869			

number	divisible by 2	divisible by 5	divisible by 10
870	x	x	x
871			
872	x		
873			
874	x		

If a number is divisible by 10, it ends in zero, so it is ALSO divisible by <u>2</u> and <u>5</u>.

Divisibility, cont.

8. a. 2, <u>4</u>, 6, <u>8</u>, 10, <u>12</u>, 14, <u>16</u>, 18, <u>20</u>, 22, <u>24</u>, 26, <u>28</u>, 30, <u>32</u>, 34, <u>36</u>, 38, <u>40</u>, 42, <u>44</u>, 46, <u>48</u>, 50, <u>52</u>, 54, <u>56</u>, 58, <u>60</u>
This is also a list of multiples of (or multiplication table of) 2.

   b. 2, <u>4</u>, 6, <u>8</u>, 10, <u>12</u>, 14, <u>16</u>, 18, <u>20</u>, 22, <u>24</u>, 26, <u>28</u>, 30, <u>32</u>, 34, <u>36</u>, 38, <u>40</u>, 42, <u>44</u>, 46, <u>48</u>, 50, <u>52</u>, 54, <u>56</u>, 58, <u>60</u>
These are every other number in the list of multiples of 2.

   c. 2, <u>4</u>, 6, <u>8</u>, 10, <u>12</u>, 14, <u>16</u>, 18, <u>20</u>, 22, <u>24</u>, 26, <u>28</u>, 30, <u>32</u>, 34, <u>36</u>, 38, <u>40</u>, 42, <u>44</u>, 46, <u>48</u>, 50, <u>52</u>, 54, <u>56</u>, 58, <u>60</u>
These are every third number in the list of multiples of 2, or every third even number divisible by 6.

   d. 12, 24, 36, 48, and 60 - or multiples of 12.

9. a. 3, 6, 9, 12, 15, 18, 21, 24, 27, 30, 33, 36, 39, 42, 45, 48, 51, 54, 57, 60
This is also a list of multiples of (or multiplication table of) 3.

   b. 3, <u>6</u>, 9, <u>12</u>, 15, <u>18</u>, 21, <u>24</u>, 27, <u>30</u>, 33, <u>36</u>, 39, <u>42</u>, 45, <u>48</u>, 51, <u>54</u>, 57, <u>60</u>
These are every second number in the list of multiples of 3.

   c. 3, <u>6</u>, 9, <u>12</u>, 15, <u>18</u>, 21, <u>24</u>, 27, <u>30</u>, 33, <u>36</u>, 39, <u>42</u>, 45, <u>48</u>, 51, <u>54</u>, 57, <u>60</u>
These are every third number in the list of multiples of 3.

10. 18, 36, 54

11. 1

12. It is also a multiple of 1, 2, 10, and 20.

Mystery number: 33 and 60

Divisibility and Factors, pp. 11-13

1. Divide. For example, divide 623 by 7 and see if the division is even. If yes, 7 is a factor of 623, otherwise not.

2. Justifications will vary. Check the student's justifications.

   a. No, because 100 ÷ 8 is not an even division.
   b. No, because 9,896 ÷ 7 leaves a remainder.
   c. No, because 50 ÷ 9 is not an even division.

3. He is not correct. Using long division, we get that 607 ÷ 13 = 46 R9. This is not an even division, so 607 is not divisible by 13. (Mark seems to be confusing the divisibility test by 3 with divisibility by 13.)

4. a. It is divisible by 3, because it is 3 times something; namely, it is 3 × (4 × 87).
   b. Yes, because 3 × 4 × 87 = 12 × 87. Since this number is 12 times some number, it is divisible by 12.
   c. Yes, because 2 × 5 equals 10, so this number is actually (10 × 758) or 7,580, which naturally is divisible by 10.

5.

Divisible by	2	5	10	100	1000
825		X			
400	X	X	X	X	
332	X				

Divisible by	2	5	10	100	1000
600,200	X	X	X	X	
56,000	X	X	X	X	X
307,995		X			

6. Explanations vary.
   a. No, because 6 × 28 ends in 8. (Imagine multiplying in columns. The first multiplication is 6 × 8 = 48, which makes the answer end in 8.)
   b. No, because 3 × 794 ends in 2. (Imagine multiplying in columns. The first multiplication is 3 × 4 = 12, which makes the entire answer to end in 2.)
   c. No. When you multiply odd numbers, the product is also odd, so it cannot be divisible by 2.
   d. Yes. Since 2 × 2 = 4, this number is (4 × 15 × 7), or 4 times some number. So, the number must be divisible by 4.

Divisibility and Factors, cont.

7. a. No.   b. Yes. $43,719 \div 3 = 14,573$ c. No

8. Change the last 2 to a 1 or to a 7. We get $238,881 \div 3 = 79,627$ or $238,887 \div 3 = 79,629$.

9. a. No   b. Yes. $576 \div 9 = 64$ c. Yes. $44,082 \div 9 = 4,898$.

10.

Divisible by	2	3	5	6	9
589					
558	X	X		X	X

Divisible by	2	3	5	6	9
495		X	X		X
3,594	X	X		X	

More on Divisibility, pp. 14-15

1. 686, 679, 672, 665, 658

2.

a. $26 \div 4 = 6$ R2	b. $78 \div 3 = 26$ R0	c. $54 \div 7 = 7$ R5
$27 \div 4 = 6$ R3	$79 \div 3 = 26$ R1	$55 \div 7 = 7$ R6
$28 \div 4 = 7$ R0	$80 \div 3 = 26$ R2	$56 \div 7 = 8$ R0
$29 \div 4 = 7$ R1	$81 \div 3 = 27$ R0	$57 \div 7 = 8$ R1
$30 \div 4 = 7$ R2	$82 \div 3 = 27$ R1	$58 \div 7 = 8$ R2
$31 \div 4 = 7$ R3	$83 \div 3 = 27$ R2	$59 \div 7 = 8$ R3
$32 \div 4 = 8$ R0	$84 \div 3 = 28$ R0	$60 \div 7 = 8$ R4

3. a. The remainder is 1.   b. The remainder is 3.

4. The remainder is 5.

5. 540

6.

Divisible by	2	3	4	5	6	9
1,755		X		X		X
298	X					
4,000	X		X	X		
3,270	X	X		X	X	

Divisible by	2	3	4	5	6	9
3,548	X		X			
277						
237		X				
10,999						

7. a. 91 or 97   b. 37

8.

18	52	100	502	300	312	348	322
16	44	64	446	292	144	360	422
6	16	72	292	280	266	436	232
86	94	104	144	216	204	568	522
60	54	128	132	244	286	572	588
12	8	12	90	308	312	78	544
15	12	136	98	254	308	348	548
44	48	66	166	256	388	428	444

9. Mystery numbers: a. 32   b. 210   c. 84   d. 132

Prime Numbers, pp. 16-18

1.

number	divisible by 1	divisible by 2	divisible by 3	divisible by 4	divisible by 5	divisible by 6	divisible by 7	divisible by 8	divisible by 9	divisible by 10
2	x	x								
3	x		x							
4	x	x		x						
5	x				x					
6	x	x	x			x				
7	x						x			
8	x	x		x				x		
9	x		x						x	
10	x	x			x					x
11	x									
12	x	x	x	x		x				
13	x									
14	x	x					x			
15	x		x		x					
16	x	x		x				x		
17	x									
18	x	x	x			x			x	
19	x									
20	x	x		x	x					x
21	x		x				x			
22	x	x								
23	x									
24	x	x	x	x		x		x		
25	x				x					
26	x	x								
27	x		x						x	
28	x	x		x			x			
29	x									
30	x	x	x		x	x				x
31	x									
32	x	x		x				x		
33	x		x							
34	x	x								
35	x				x		x			

2. Prime numbers: 2, 3, 5, 7, 11, 13, 17, 19, 23, 29, 31

3. Answers will vary, as you can write a composite number as a product in many different ways.

a. 33 is composite. $33 = 3 \times 11$	b. 52 is composite. $52 = 2 \times 26$	c. 41 is prime.
d. 39 is composite. $39 = 3 \times 13$	e. 43 is prime.	f. 45 is composite. $45 = 5 \times 9$

Prime Numbers, cont.

4.

number	digit sum	divisible by 3?
98	17	no
105	6	yes
567	18	yes
59	14	no

number	digit sum	divisible by 3?
888	24	yes
1,045	10	no
1,338	15	yes
612	9	yes

5.

number	divisible by 7?
99	no
74	no
56	yes

number	divisible by 7?
24	no
100	no
84	yes

number	divisible by 7?
85	no
63	yes
105	yes

6. Answers will vary, as you can write a composite number as a product in many different ways.

a. 67 is prime.	b. 57 is composite. $57 = 3 \times 19$	c. 47 is prime.
d. 53 is prime.	e. 63 is composite. $63 = 7 \times 9$	f. 61 is prime.
g. 93 is composite. $93 = 3 \times 31$	h. 85 is composite. $85 = 5 \times 17$	i. 91 is composite. $91 = 7 \times 13$
j. 87 is composite. $87 = 3 \times 29$	k. 79 is prime.	l. 97 is prime.

Finding Factors, pp. 19-20

1.

a. factors: 1, 2, 3, 6	b. factors: 1, 2, 5, 10
c. factors: 1, 2, 3, 4, 6, 12	d. factors: 1, 3, 5, 15
e. factors: 1, 2, 4, 5, 10, 20	f. factors: 1, 2, 3, 6, 9, 18

2. Only Olivia's work was totally correct.

a. Aiden found all the factors of 34: ~~$34 = 2 \times 18$~~ $34 = 2 \times 17$ ~~$34 = 1 \times 17$~~ $34 = 1 \times 34$ The factors are 1, 2, 17, ~~18~~, 34	b. Olivia found all the factors of 28: $28 = 1 \times 28$ $28 = 2 \times 14$ $28 = 4 \times 7$ The factors are 1, 2, 4, 7, 14, and 28.
c. Jayden found all the factors of 33: $33 = 1 \times 33$ ~~$33 = 3 \times 13$~~ $33 = 3 \times 11$ The factors are 1, 3, ~~13~~, 11, 33.	d. Isabella found all the factors of 36: $36 = 6 \times 6$ <u>$36 = 3 \times 12$</u> <u>$36 = 3 \times 12$</u> $36 = 4 \times 9$ <u>$36 = 1 \times 36$</u> The factors are 4, 6, and 9. <u>Also 1, 2, 3, 12, 18, 36</u>

Finding Factors, cont.

3.

a. factors: 1, 2, 23, 46	b. factors: 1, 2, 4, 17, 34, 68
c. factors: 1, 3, 9, 11, 33, 99	d. factors: 1, 2, 3, 4, 6, 8, 9, 12, 18, 24, 36, 72
e. factors: 1, 73	f. factors: 1, 2, 4, 5, 8, 10, 16, 20, 40, 80
g. factors: 1, 5, 19, 95	h. factors: 1, 2, 4, 8, 16, 32, 64

Primes and Finding Factors, pp. 21-23

1. a. 1, 2, 13, 26   b. 1, 2, 19, 38
   c. 1, 2, 4, 8, 11, 22, 44, 88   d. 1, 47
   e. 1, 71   f. 1, 2, 43, 86

2.

Number	Factors
11	1, 11
12	1, 2, 3, 4, 6, 12
13	1, 13
14	1, 2, 7, 14
15	1, 3, 5, 15
16	1, 2, 4, 8, 16
17	1, 17
18	1, 2, 3, 6, 9, 18
19	1, 19
20	1, 2, 4, 5, 10, 20

3. List of primes between 1 and 20: 2, 3, 5, 7, 11, 13, 17, 19

5. a. 53 or 59   b. 83 or 89

Puzzle corner. a. 113   b. 36   c. 30

Prime Factorization, pp. 24-28

Examples of factoring some composite numbers: $57 = 3 \times 19$.

1. a. $18 = 2 \times 3 \times 3$ b. $6 = 2 \times 3$ c. $14 = 2 \times 7$
   d. $8 = 2 \times 2 \times 2$ e. $12 = 2 \times 2 \times 3$ f. $20 = 2 \times 2 \times 5$
   g. $16 = 2 \times 2 \times 2 \times 2$ h. $24 = 2 \times 2 \times 2 \times 3$ i. $27 = 3 \times 3 \times 3$
   j. $25 = 5 \times 5$ k. $33 = 3 \times 11$ l. $15 = 3 \times 5$

2. a. $42 = 2 \times 3 \times 7$ b. $56 = 2 \times 2 \times 2 \times 7$ c. $68 = 2 \times 2 \times 17$
   d. $75 = 3 \times 5 \times 5$ e. $47 = 1 \times 47$ f. $99 = 3 \times 3 \times 11$
   g. $72 = 2 \times 2 \times 2 \times 3 \times 3$ h. $80 = 2 \times 2 \times 2 \times 2 \times 5$ i. $97 = 1 \times 97$
   j. $85 = 5 \times 17$ k. $66 = 2 \times 3 \times 11$ l. $82 = 2 \times 41$

3. a. 110   b. 24   c. 42   d. 66   e. 90   f. 102

Prime Factorization, cont.

4. a. 130   b. 2,002   c. 570

5. Answers will vary. Please check the students' work.

Puzzle corner: a. $2{,}145 = 3 \times 5 \times 11 \times 13$ b. $3{,}680 = 2 \times 2 \times 2 \times 2 \times 2 \times 5 \times 23$ c. $10{,}164 = 2 \times 2 \times 3 \times 7 \times 11 \times 11$

The Sieve of Eratosthenes and Prime Factorization, pp. 29-32

Primes between 0 and 100: 2, 3, 5, 7, 11, 13, 17, 19, 23, 29, 31, 37, 41, 43, 47, 53, 59, 61, 67, 71, 73, 79, 83, 89, 97

1. a. Factors of 54 are 1, 2, 3, 6, 9, 18, 27, 54
   b. Factors of 60 are 1, 2, 3, 4, 5, 6, 10, 12, 15, 20, 30, 60
   c. Factors of 84 are 1, 2, 3, 4, 6, 7, 12, 14, 21, 28, 42, 84
   d. Factors of 97 are 1, 97 (it is prime).

2.

a. $124 = 2^2 \cdot 31$	b. $260 = 2^2 \cdot 5 \cdot 13$	c. $96 = 2^5 \cdot 3$
d. $90 = 2 \cdot 3^2 \cdot 5$	e. $165 = 3 \cdot 5 \cdot 11$	f. $95 = 5 \cdot 19$

3.

Divisible by	2	3	4	5	6	9
128	x		x			
765		x		x		x

Divisible by	2	3	4	5	6	9
209						
6,042	x	x			x	

4.

	a. $912 = 2^4 \cdot 3 \cdot 19$	b. $528 = 2^4 \cdot 3 \cdot 11$
c. $76 = 2^2 \cdot 19$	d. $126 = 2 \cdot 3^2 \cdot 7$	e. $272 = 2^4 \cdot 17$

5. The factorization of 164 is wrong. In the factor tree, the second line shows $4 \cdot 42$. This should be $4 \cdot 41$. In reality, the prime factorization of 164 is $2 \cdot 2 \cdot 41$.

6. The odd numbers to consider are 101, 103, 105, 107, and 109. Of those, 105 is divisible by 5 and by 3. None of the others are divisible by 3 or 9. Thus, we need to check divisibility by 7. Since $98 = 7 \cdot 14$, then the next number divisible by 7 is 105. Thus, 101, 103, 107, and 109 are all primes.

The Sieve of Eratosthenes and Prime Factorization, cont.

7.

a. $196 = 2^2 \cdot 7^2$	b. $380 = 2^2 \cdot 5 \cdot 19$	c. $336 = 2^4 \cdot 3 \cdot 7$
d. $306 = 2 \cdot 3^2 \cdot 17$	e. $116 = 2^2 \cdot 29$	f. $720 = 2^4 \cdot 3^2 \cdot 5$
g. $675 = 3^3 \cdot 5^2$	h. $990 = 2 \cdot 3^2 \cdot 5 \cdot 11$	i. $945 = 3^3 \cdot 5 \cdot 7$

Puzzle corner:
2, 3, 5, 7, 11, 13, 17, 19, 23, 29, 31, 37, 41, 43, 47, 53, 59, 61, 67, 71, 73,
79, 83, 89, 97, 101, 103, 107, 109, 113, 127, 131, 137, 139, 149, 151,
157, 163, 167, 173, 179, 181, 191, 193, 197, 199

The Greatest Common Factor (GCF), pp. 33-35

1. a. 1, 2, 3, 4, 6, 8, 12, 16, 24, 48   b. 1, 2, 3, 4, 5, 6, 10, 12, 15, 20, 30, 60
   c. 1, 2, 3, 6, 7, 14, 21, 42   d. 1, 3, 9, 11, 33, 99

2. a. 12   b. 6   c. 6   d. 3

3. a. 1, 2, 4, 11, 22, 44   b. 1, 2, 3, 6, 11, 22, 33, 66
   c. 1, 2, 4, 7, 14, 28   d. 1, 2, 4, 7, 8, 14, 28, 56
   e. 1, 2, 4, 5, 10, 20, 25, 50, 100   f. 1, 3, 5, 9, 15, 45

4. a. 22   b. 4   c. 5   d. 3
   e. 4   f. 28   g. 4   h. 1

5. a. $\dfrac{8}{11}$ b. $\dfrac{21}{22}$ c. $\dfrac{7}{8}$ d. $1\dfrac{13}{20}$

   e. $\dfrac{12}{25}$ f. $1\dfrac{1}{99}$ g. 2   h. $\dfrac{4}{9}$

6. a. 6   b. 4   c. 2
   d. 12   e. 1   f. 8

7. a. 5   b. 2
   c. 6   d. 12

Puzzle corner: Since $11 \cdot 17 = 187$ and $11 \cdot 24 = 264$, the greatest common factor of 187 and 264 is 11.

Factoring Sums, pp. 36-38

1.

a. $3(x + 6)$ and $3 \cdot x + 3 \cdot 6 = 3x + 18$	b. $8(11 + b)$ and $8 \cdot 11 + 8 \cdot b = 88 + 8b$
c. $8(11 + 20)$ and $8 \cdot 11 + 8 \cdot 20 = 248$	d. $7(9 + 11)$ and $7 \cdot 9 + 7 \cdot 11 = 140$

2. 1 cm × 30 cm, 2 cm × 15 cm, 3 cm × 10 cm and 5 cm × 6 cm

3. 1 m × 40 m, 2 m × 20 m, 4 m × 10 m, and 5 m × 8 m.

4. Answers will vary. The answer here shows a length of 5 cm for the shared side.

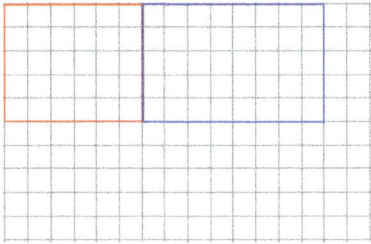

5. The sides of 1 cm, 2 cm and 5 cm can all be used for the answer in problem 4. So, the rectangles that are side-by-side can be 30 cm × 1 cm and 40 cm × 1 cm, OR 15 cm × 2 cm and 20 cm × 2 cm, OR 6 cm × 5 cm and 8 cm × 5 cm.

6.

a. The GCF of 18 and 12 is **6** . $18 + 12 = $ **6** $\cdot\ 3\ +$ **6** $\cdot\ 2\ =$ **6** $(3 + 2)$
b. The GCF of 6 and 10 is 2. $6 + 10 = 2 \cdot 3 + 2 \cdot 5 = 2(3 + 5)$
c. The GCF of 22 and 11 is 11. $22 + 11 = 11 \cdot 2 + 11 \cdot 1 = 11(2 + 1)$

6. (continued)

d. The GCF of 15 and 21 is 3. $15 + 21 = 3 \cdot 5 + 3 \cdot 7 = 3(5 + 7)$
e. The GCF of 25 and 35 is 5. $25 + 35 = 5(5 + 7)$
f. The GCF of 72 and 86 is 2. $72 + 86 = 2\ (36 + 43)$
g. The GCF of 96 and 40 is 8. $96 + 40 = 8(12 + 5)$
h. The GCF of 39 and 81 is 3. $39 + 81 = 3(13 + 27)$

7. a. $8(4 + 5)$ b.

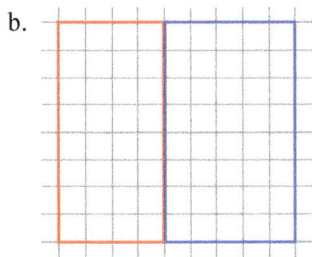

8. The rectangles are 6 × 5 and 5 × 5:

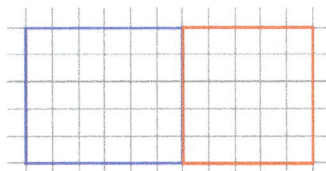

Factoring Sums, cont.

9. See the image on the right. The rectangles are 7×6, 4×6 and 5×6.

10. Since $45 = 3 \cdot 3 \cdot 5$, the pen can be 9 m $\times$ 5 m, 3 m $\times$ 15 m, or 45 m $\times$ 1 m. (The last one isn't a very likely size).

11. a. Answers will vary. Please check the student's answers. For example, 3 and 4, or 15 and 16. There is an infinite number of possible correct answers.

    b. Answers will vary. Please check the student's answers. For example, 89 and 88, or 26 and 37.

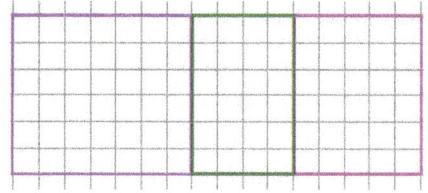

12.

a. The GCF of 15 and 5 is 5. $15x + 5 = 5(3x + 1)$	b. The GCF of 18 and 30 is 6. $18x + 30 = 6(3x + 5)$
c. The GCF of 72 and 54 is 9. $72a + 54b = 9(8a + 6b)$	d. The GCF of 100 and 90 is 10. $100y + 90x = 10(10y + 9x)$

Puzzle corner:
a. Answers will vary. Please check the student's answers. Example: 26 and 39.
b. Answers will vary. Please check the student's answers. Example: 51 and 204.

The Least Common Multiple (LCM), pp. 39-42

1. a. 6   b. 18
   c. 56   d. 24
   e. 70   f. 30

2. a. Answers will vary. Please check the student's answers.
      Any four of these numbers will work: 6, 12, 18, 24, 30, 36, 42, 48, 54, 60, 66, 72, 78, 84, 90 or 96.

   b. 96
   c. 1,250

3. The LCM of 12 and 15 is 60. This means the next time there is a bus for both routes is in 60 minutes, or at 4:30 PM.

4. The LCM of 20 and 45 is 180. At the height of 180 cm, both stacks are the same height.

5. a. no   b. no
   c. no   d. yes
   e. yes   f. no

Example 4. The LCM of 8 and 10 is 40.

6. a. 12   b. 63
   c. 10   d. 28
   e. 10   f. 20

7. a.

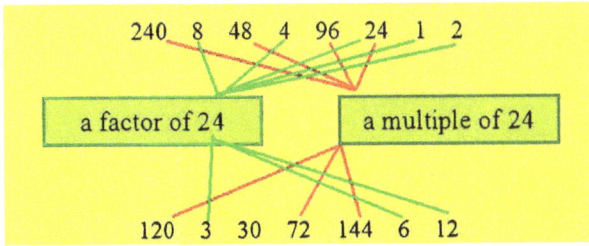

b. The number 30 is neither a multiple nor a factor of 24.
c. The number 24 is BOTH a factor and a multiple of 24.

8. a. 24   b. 30   c. 30   d. 56

9. The LCM of 8, 6, and 4 is 24. So, they will have a haircut again on the same day in <u>24 weeks</u>.

10. a. $\frac{17}{72}$ b. $\frac{23}{24}$ c. $\frac{9}{70}$ d. $\frac{13}{18}$

11. a. 120   b. 120   c. 80   d. 600

12.

a. $\frac{29}{40}$ and $\frac{11}{15}$	b. $\frac{11}{20}$ and $\frac{13}{24}$	c. $\frac{7}{20}$ and $\frac{5}{16}$	d. $\frac{39}{50}$ and $\frac{94}{120}$
$\downarrow$ $\downarrow$	$\downarrow$ $\downarrow$	$\downarrow$ $\downarrow$	$\downarrow$ $\downarrow$
$\frac{87}{120} < \frac{88}{120}$	$\frac{66}{120} > \frac{65}{120}$	$\frac{28}{80} > \frac{25}{80}$	$\frac{468}{600} < \frac{470}{600}$

Puzzle corner: <u>Day number 236 is Saturday.</u> We need to divide 236 by seven, first of all, to find out how many complete weeks have passed. We also need take a careful look at the remainder to find how many additional days have passed. $236 \div 7 = 33$ R5   So, 33 complete weeks and 5 additional days have passed. The 33 completed weeks would take us to a certain Monday. $33 \cdot 7 = 231$ days. So, day 231 is a Monday. Then:

day 232 - Tuesday   day 235 - Friday
day 233 - Wednesday   day 236 - Saturday
day 234 - Thursday

Review, pp. 43-46

1.

a. 3 · 3 · 3 · 3	b. 2 · 13	c. 5 · 13
d. 3 · 2 · 2 · 2 · 2 · 2	e. 2 · 2 · 31	f. 2 · 3 · 3 · 5 · 5

2.

a. $\dfrac{28}{84} = \dfrac{\overset{1}{\cancel{4}} \cdot \overset{1}{\cancel{7}}}{\underset{3}{\cancel{21}} \cdot \underset{1}{\cancel{4}}} = \dfrac{1}{3}$

b. $\dfrac{75}{160} = \dfrac{\overset{1}{\cancel{5}} \cdot 15}{\underset{2}{\cancel{10}} \cdot 16} = \dfrac{15}{32}$

c. $\dfrac{222}{36} = \dfrac{\overset{1}{\cancel{6}} \cdot 37}{\underset{1}{\cancel{6}} \cdot \cancel{6}} = \dfrac{37}{6} = 6\dfrac{1}{6}$

d. $\dfrac{48}{120} = \dfrac{\overset{1}{\cancel{6}} \cdot \overset{4}{\cancel{8}}}{\underset{2}{\cancel{12}} \cdot \underset{5}{\cancel{10}}} = \dfrac{4}{10} = \dfrac{2}{5}$

3. a. 21   b. 40
   c. 66   d. 24

4. a. 8   b. 25
   c. 16   d. 6

5. a. 25, 50, 75, 100, 125 and 150 are <u>multiples</u> of 25.
   1, 2, 5, 10, 25 and 50 are <u>factors</u> of 50.
   Each number has an infinite number of <u>multiples</u>.
   Each number has a greatest <u>factor</u>.
   If the number x divides into another number y, we say x is a <u>factor</u> of y.

   b. Answers will vary. Please check the student's answers.
   Any five of these will work: 75, 90, 105, 120, 135, 150, 165, 180, 195.

   c. Any number that is divisible by 28. Answers will vary. Please check the student's answers.
   Example: 28, 56, 112, 224, 448. The LCM of 4 and 7 is 28.

6. a. GCF of 12 and 21 is 3.
   $12 + 21 = 3 \cdot 4 + 3 \cdot 7 = 3(4 + 7)$
   b. GCF of 45 and 70 is 5.
   $45 + 70 = 5(9 + 14)$

7.

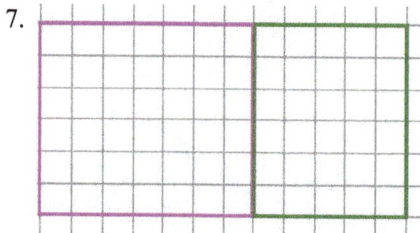

More from math MAMMOTH

Math Mammoth has a variety of resources to fit your needs. All are available as economical downloads, and most also as printed copies.

- **Math Mammoth Light Blue Series**
  A complete curriculum for grades 1-7. Each grade level includes two student worktexts (A and B), which contain all the instruction and exercises all in the same book, answer keys, tests, cumulative reviews, and a worksheet maker. International (all metric), Canadian, and South African versions are also available.

  https://www.MathMammoth.com/complete-curriculum

  https://www.MathMammoth.com/international/international

  https://www.MathMammoth.com/canada/

  https://www.MathMammoth.com/south_africa/

- **Math Mammoth Skills Review Workbooks**
  These workbooks are intended to be used alongside the Light Blue series full curriculum, and they provide additional review to the topics studied in the main curriculum, in a spiral manner.
  https://www.MathMammoth.com/skills_review_workbooks/

- **Math Mammoth Blue Series**
  Blue Series books are topical worktexts for grades 1-7, containing both instruction and exercises. The topics cover all elementary mathematics from 1st through 7th grade. These books are not tied to grade levels, and are thus great for filling in gaps.
  https://www.MathMammoth.com/blue-series

- **Make It Real Learning**
  These activity workbooks concentrate on answering the question, "Where is math used in real life?" The series includes various workbooks for grades 3-12.
  https://www.MathMammoth.com/worksheets/mirl/

- **Review Workbooks**
  Workbooks for grades 1-7 that provide a comprehensive review of one grade level of math—for example, for review during school break or summer vacation.
  https://www.MathMammoth.com/review_workbooks/

Free gift!

- Receive over 350 free sample pages and worksheets from my books, plus other freebies:
  https://www.MathMammoth.com/worksheets/free

Lastly...

- Inspire4 is an inspirational website for the whole family I've been privileged to help with:
  https://www.inspire4.com